THE SURVIVORS FIELD GUIDE TO THE BAOFENG RADIO

A GUERRILLA'S HANDBOOK ON HOW TO PROGRAM AMATEUR RADIOS FOR BEGINNERS

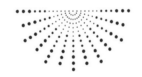

MIKE COMMANDOS

Copyright © 2023 by Mike Commandos. All rights reserved.

No part of this book may be reproduced, stored in a retrieval system, or transmitted in any form or by any means, electronic, mechanical, photocopying, recording, or otherwise, without written permission from the publisher.

This book is protected under U.S. copyright law, which states that copyright protection subsists from the time the work is created in fixed form. The copyright in the work of authorship immediately becomes the property of the author who created the work.

Published by Mike Commandos.

Printed in the United States of America.

Disclaimer: This book is not intended as a substitute for professional advice. The information provided in this book is for general informational purposes only and does not establish a professional relationship. The author and publisher make no representation or warranties with respect to the accuracy or completeness of the contents of this book and specifically disclaim any implied warranties of merchantability or fitness for a particular purpose. The author and publisher shall not be liable in any event for incidental or consequential damages in connection with or arising out of, the furnishing, performance, or use of the information in this book.

"Dedicated to connecting people and communities through the power of radio communications."

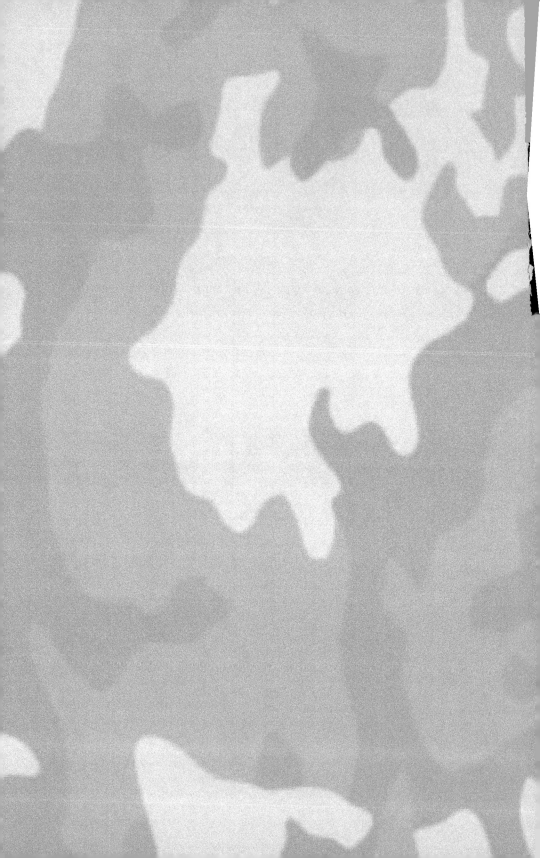

1
AN INTRODUCTION TO BAOFENG RADIO: UNDERSTANDING THE BENEFITS AND KEY FEATURES

The Baofeng Radio is a compact and versatile two-way radio that has taken the world of amateur radio enthusiasts, outdoor adventurers, and preppers by storm. This guide will provide a comprehensive overview of the radio, its key features and benefits, and help you understand why it's the perfect choice for anyone looking for reliable communication in a variety of environments.

From its user-friendly design to its wide range of frequency options, the Baofeng Radio has been designed with the user in mind. Despite its compact size, the radio packs a powerful punch, offering crystal-clear audio and robust connectivity even in remote locations. And, with its intuitive interface, even those with limited experience can easily program and operate the radio.

But the Baofeng Radio is much more than just a two-way radio. It's a versatile communication tool that offers a range of useful features, including multiple channels, the ability to store and access multiple frequencies, and the option to select from different frequency bands.

Whether you're a seasoned amateur radio operator or just starting out, the Baofeng Radio is the ideal choice for anyone looking for a reliable and flexible communication device. In the following chapters, we'll take a closer look at the different frequency options available, provide a step-by-step guide to setting up the radio, and cover key topics such as programming the channels and frequencies, optimizing the radio's settings for your specific needs, and troubleshooting common issues.

So, whether you're looking to stay connected with friends and family while camping or hiking, or you're a prepper preparing for a worst-case scenario, the Baofeng Radio has got you covered. So, let's get started and explore all that this amazing radio has to offer!

FROM THE OFFICIAL USER'S MANUAL OF THE BAOFENG RADIO:

The Baofeng radio configuration information in the official user's manual typically includes the following:

1. Power On/Off: Instructions for turning the radio on and off, as well as how to conserve battery life.
2. Channel Programming: A step-by-step guide on how to program channels, including the frequency, channel number, and channel name.
3. Tone Settings: Information on how to set the tone for a specific channel, including the tone frequency and tone type (CTCSS or DCS).

4. **Duplex Settings:** Explanation of duplex settings, including how to set the transmit and receive frequencies for repeater operation.
5. **Scan Settings:** Detailed instructions on how to scan channels, including how to enable and disable the scan function.
6. **Battery and Charging:** Information on how to charge the radio battery, as well as tips for maximizing battery life.
7. **Antenna Settings:** Explanation of the different types of antennas that can be used with the Baofeng radio, including the proper way to attach and detach the antenna.
8. **Display Settings:** Information on how to adjust the display settings, including the backlight and contrast settings.
9. **Sound Settings:** Details on how to adjust the volume, as well as how to set the tone for incoming and outgoing calls.
10. **Radio Settings:** Information on how to change the default settings, including the time and date, language, and power saving mode.
11. **Troubleshooting:** Common problems that users may encounter and how to resolve them, including no sound, poor reception, and battery issues.

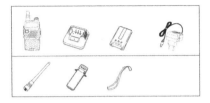

1. Technical Specifications: A complete list of the technical specifications for the Baofeng radio, including frequency range, power output, and dimensions.

Here are practical examples for each of the 12 points listed in the Baofeng radio configuration procedure:

1. **Setting the Frequency:** To set the frequency, you first need to turn on your radio and select the frequency mode (VFO or MR). Then, using the up and down buttons, adjust the frequency to your desired channel. For example, if you want to set the frequency to 146.520 MHz, you will press the up button until the radio displays 146.520.
2. **Setting the Tone:** To set the tone, you need to first select the tone mode (CTCSS or DCS). Then, using the up and down buttons, adjust the tone frequency to the desired setting. For

example, if you want to set the tone to 100.0 Hz, you will press the up button until the radio displays 100.0.

3. **Setting the Channel Name:** To set the channel name, you need to first enter the channel programming mode. Then, using the up and down buttons, select the channel you want to rename. Finally, use the keypad to input the desired name for the channel. For example, if you want to name channel 1 "Local Repeater," you would press the keypad buttons to enter the letters "L-O-C-A-L-R-E-P".

4. **Setting the Step Size:** To set the step size, you need to first enter the menu mode. Then, using the up and down buttons, select "Step Size" and adjust the step size to your desired setting. For example, if you want to set the step size to 5 KHz, you will press the up button until the radio displays "5.0 KHz".

5. **Setting the Battery Save:** To set the battery save, you need to first enter the menu mode. Then, using the up and down buttons, select "Battery Save" and adjust the setting to either "On" or "Off". For example, if you want to turn the battery save on, you will press the up button until the radio displays "On".

6. **Setting the Timeout Timer:** To set the timeout timer, you need to first enter the menu mode.

Then, using the up and down buttons, select "Timeout Timer" and adjust the setting to your desired time in minutes. For example, if you want to set the timeout timer to 5 minutes, you will press the up button until the radio displays "5".

7. **Setting the Keypad Tone:** To set the keypad tone, you need to first enter the menu mode. Then, using the up and down buttons, select "Keypad Tone" and adjust the setting to either "On" or "Off". For example, if you want to turn the keypad tone on, you will press the up button until the radio displays "On".

8. **Setting the VOX:** To set the VOX, you need to first enter the menu mode. Then, using the up and down buttons, select "VOX" and adjust the setting to either "On" or "Off". For example, if you want to turn VOX on, you will press the up button until the radio displays "On".

9. **Setting the Power Output:** To set the power output, you need to first enter the menu mode. Then, using the up and down buttons, select "Power Output" and adjust the setting to your desired level (High, Medium, or Low). For example, if you want to set the power output to high, you will press the up button until the radio displays "High".

10. **Setting the Squelch Level:** To set the squelch level on the Baofeng radio, follow these steps:

- Turn the radio on and make sure it is in the "frequency mode".
- Locate the "SQL" button on the side of the radio and press it until the squelch level display appears on the screen.
- Turn the dial or use the up and down arrows to adjust the squelch level until the background noise is eliminated. The squelch level should be set just high enough to eliminate the noise, but not too high to interfere with the incoming signal.
- Once the desired squelch level is set, press the "SQL" button to return to frequency mode.
- Test the squelch level by turning the radio to a weak or noisy channel and verifying that the noise is eliminated when the signal is too weak.
- It is important to note that the squelch level will need to be adjusted for different environments and frequencies, so it is a good idea to familiarize yourself with this feature and make adjustments as needed.

1. **Setting the Dual Watch Function:** Dual Watch is a feature that allows you to monitor two different frequencies simultaneously. To set up

Dual Watch on your Baofeng radio, follow these steps:

- Press the Menu button to access the main menu
- Use the channel up/down buttons to navigate to the "Dual Watch" option
- Press the Menu button to select the "Dual Watch" option
- Use the channel up/down buttons to select either "On" or "Off" for the Dual Watch function
- Press the Menu button to save the changes and exit the menu
- Example: Let's say you want to monitor channel 1 (146.520 MHz) and channel 2 (147.540 MHz) at the same time. To do this, you would navigate to the "Dual Watch" option in the menu, select "On," and then save the changes. Now, your Baofeng radio will display both channels 1 and 2, and you can toggle between them to listen to each frequency.

1. **Setting the Battery Save Function:** The Battery Save function helps to conserve battery power by reducing the power output when the radio is not being used. To set up the Battery

Save function on your Baofeng radio, follow these steps:

- Press the Menu button to access the main menu
- Use the channel up/down buttons to navigate to the "Battery Save" option
- Press the Menu button to select the "Battery Save" option
- Use the channel up/down buttons to select either "On" or "Off" for the Battery Save function
- Press the Menu button to save the changes and exit the menu
- Example: If you're using your Baofeng radio in an area where battery life is critical, you can turn on the Battery Save function to extend the battery life. To do this, you would navigate to the "Battery Save" option in the menu, select "On," and then save the changes. Now, your Baofeng radio will automatically reduce its power output when it's not being used, helping to conserve battery power.

FROM THE OFFICIAL USER'S MANUAL OF THE BAOFENG ... | 11

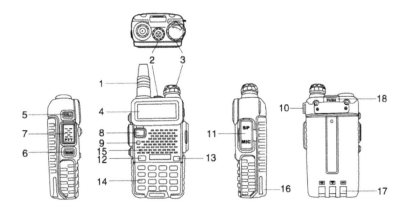

1. antenna
2. flashlight
3. knob (ON/OFF,volume)
4. LCD
5. SK-side key1/CALL(radio,alarm)
6. SK-side key2/MONI(flashlight,monitor)
7. PTT key(push-to-talk)
8. VFO/MR (frequency mode/channel mode)
9. LED indicator
10. strap buckle
11. accessory jack
12. A/B key(frequency display switches)
13. BAND key(band switches)
14. keypad
15. SP.&MIC.
16. battery pack
17. battery contacts
18. battery remove button

LCD' DISPLAY:

The display icons appear when certain operations or specific features are activated.

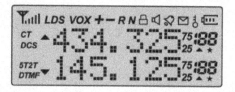

Icon	Description
:88	Operating channel.
75/25	Operating frequency.
CT	'CTCSS' activated.
DCS	'DCS' activated.
+−	Frequency offset direction for accessing repeaters.
S	Dual Watch/Dual Reception functions activated.
VOX	Function 'VOX' enabled.
R	Reverse function activated.
N	Wide Band selected.
🔋	Battery Level indicator
🔑	Keypad lock function activated.
L	Low transmit power.
▲▼	Operation frequency.
📶	Signal Strength Level.

FROM THE OFFICIAL USER'S MANUAL OF THE BAOFENG ... | 13

· 1750 Hz TONE FOR ACCESS TO REPEATERS:

The user needs to establish long distance communications through an amateur radio repeater which is activated after receiving a 1750 Hz tone. Press and hold on the [PTT],then press the [BAND] button to transmit a 1750Hz tone.

BASIC OPERATION:

·RADIO ON-OFF/VOLUME CONTROL :
-Make sure the antenna and battery are installed correctly and the battery charged.
-Rotate the knob clockwise to turn the radio on, and rotate the knob fully counter-clockwise until a 'click' is heard to turn the radio off. Turn the knob clockwise to increase the volume, or counter-clockwise to decrease the volume.

- SELECTING A FREQUENCY OR CHANNEL:
-Press the key[▲]or[▲]to select the desired frequency/channel you want. The display shows the frequency / channel selected.
-Press and hold down the key [▲]or[▲] for frequency up or down fast.

Note:
- You can not select a channel if not previously stored.

·ADVANCED OPERATION:

You can program your transceiver operating in the setup menu to suit your needs or preferences.

·SET MENU DESCRIPTION:

Menu	Function/Description	Available settings
0	SQL (Squelch level)	0-9
1	STEP(Frequency step)	2.5/5/6.25/10/12.5/25kHz
2	TXP(Transmit power)	HIGH/LOW
3	SAVE(Battery save,1:1/1:2/1:3/1:4)	OFF/1/2/3/4
4	VOX(Voice operated transmission)	OFF/0-10
5	W/N(Wideband/narrowband)	WIDE/NARR
6	ABR(Display illumination)	OFF/1/2/3/4/5s
7	TDR(Dual watch/dual reception)	OFF/ON

8	BEEP(Keypad beep)	OFF/ON
9	TOT(Transmission timer)	15/30/45/60.../585/600seconds
10	R-DCS(Reception digital coded squelch)	OFF/D023N...D754I
11	R-CTS(Reception Continuous Tone Coded Squelch)	67.0Hz...254.1Hz
12	T-DCS(Transmission digital coded squelch)	OFF/D023N...D754I
13	T-CTS(Transmission Continuous Tone Coded Squelch)	67.0Hz...254.1Hz
14	VOICE(Voice prompt)	OFF/ON
15	ANI(Automatic number identification of the radio,only can be set by PC software.	
16	DTMFST(The DTMF tone of transmitting code.)	OFF/DT-ST/ANI-ST/DT+ANI
17	S-CODE(Signal code, only could be set by PC software.)	1,...,15 groups
18	SC-REV(Scan resume method)	TO/CO/SE
19	PTT-ID(press or release the PTT button to transmit the signal code)	OFF/BOT/EOT/BOTH
20	PTT-LT(delay the signal code sending)	0,...,30ms
21	MDF-A(under channel mode, A channel displays. Note: name display only can be set by PC software.	FREQ/CH/NAME
22	MDF-B(under channel mode, B channel displays. Note: name display only can be set by PC software.	FREQ/CH/NAME
23	BCL(busy channel lockout)	OFF/ON
24	AUTOLK(keypad locked automatically)	OFF/ON
25	SFT-D(direction of frequency shift)	OFF/+/-
26	OFFSET(frequency shift)	00.000...69.990
27	MEMCH(stored in memory channels)	000, ...127
28	DELCH(delete the memory channels)	000, ...127
29	WT-LED(illumination display color of standby)	OFF/BLUE/ORANGE/PURPLE
30	RX-LED(illumination display color of reception)	OFF/BLUE/ORANGE/PURPLE
31	TX-LED(illumination display color of transmitting)	OFF/BLUE/ORANGE/PURPLE
32	AL-MOD(alarm mode)	SITE/TONE/CODE
33	BAND(band selection)	VHF/UHF
34	TX-AB(transmitting selection while in dual watch/ reception)	OFF/A/B
35	STE(Tail Tone Elimination)	OFF/ON

36	RP_STE(Tail tone elimination in communication through repeater)	OFF/1,2,3...10
37	RPT_RL(Delay the tail tone of repeater)	OFF/1,2,3...10
38	PONMGS(Boot display)	FULL/MGS
39	ROGER(tone end of transmission)	ON/OFF
40	RESET (Restore to default setting)	VFO/ALL

-SHORTCUT MENU OPERATION:
1.-Press the key MENU,then press the key ▲ or ▼ to select the desired menu.
2.-Press the key MENU again, come to the parameter setting.
3.-Press the key ▲ or ▼ to select the desired parameter.
4.-Press the key MENU to confirm and save, press the key EXIT to cancel setting or clear the input.

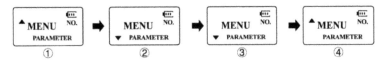

-Note:
Under channel mode,the following menu settings are invalid:CTCSS,DCS,W/N,PTT-ID,BCL,SCAN ADD TO,S-CODE,CHANNEL NAME.Only the H/L power could be changed.

-"SQL" (SQUELCH):
-The squelch mute the speaker of the transceiver in the absence of reception. With the squelch level correctly set, you will hear sound only while actually receiving signals and significantly reduces battery current consumption. It is recommended that you set Level 5.

· FUNCTION "VOX" (VOICE OPERATED TRANSMISSION):
-This function is not necessary to push the [PTT] on the transceiver for a transmission. Transmission is activated automatically by detecting the radio voice. When finish speaking, the transmission automatically terminated and the transceiver will automatically receive signal. Be sure to adjust the VOX Gain level to an appropriate sensitivity to allow smooth transmission.

· SELECT WIDEBAND OR NARROW BAND "W/N":
In areas where the RF signals are saturated, you must use the narrow band of transmission to avoid interference in adjacent channels.

- TDR (DUAL WATCH/DUAL RECEPTION):
This feature allows you to operate between frequency A and frequency B. Periodically, the transceiver checks whether a signal is received on another frequency that we have scheduled. If you receive a signal, the unit will remain in the frequency until the received signal disappears.

3
UNDERSTANDING THE DIFFERENT FREQUENCY BANDS AND THEIR USES

The Baofeng Radio offers a range of frequency bands to choose from, each with its own set of benefits and uses. In this chapter, we'll explore the different frequency bands available and help you understand which one is right for you.

First, let's start with the most commonly used frequency band: the 2-meter band. This band is used by amateur radio operators for local communication and is often referred to as the "FM simplex" band. With a range of up to 100 miles, the 2-meter band is perfect for staying connected with friends and family while out in the wilderness.

Another popular frequency band is the 70-centimeter band, also known as the "440 MHz band." This band is used for both local and regional communication and offers a range of up to 30 miles. It's a great option for preppers and outdoor enthusiasts who need reliable communication over shorter distances.

Next, let's talk about the 1.25-meter band, also known as the "220 MHz band." This band is a lesser-known option but offers a range of up to 20 miles. It's a great choice for those who need to stay connected over short distances in urban environments.

Finally, we have the 10-meter band, also known as the "28 MHz band." This band is used for long-distance communication and is often referred to as the "HF simplex" band. With a range of up to 1000 miles, the 10-meter band is perfect for those who need to stay connected over long distances, such as preppers and outdoor enthusiasts who may be traveling great distances from civilization.

In summary, Baofeng Radio offers a range of frequency bands to choose from, each with its own set of benefits and uses. Whether you're looking for local communication over short distances, or long-distance communication over great distances, Baofeng Radio has got you covered. In the following chapters, we'll delve into each frequency band in more detail and provide tips and tricks for optimizing your radio's performance.

CTCSS TABLE:

N°	Tone(Hz)	N°	Tone(Hz)	N°	Tone(Hz)	N°	Tone(Hz)	N°	Tone(Hz)
1	67.0	11	94.8	21	131.8	31	171.3	41	203.5
2	69.3	12	97.4	22	136.5	32	173.8	42	206.5
3	71.9	13	100.0	23	141.3	33	177.3	43	210.7
4	74.4	14	103.5	24	146.2	34	179.9	44	218.1
5	77.0	15	107.2	25	151.4	35	183.5	45	225.7
6	79.7	16	110.9	26	156.7	36	186.2	46	229.1
7	82.5	17	114.8	27	159.8	37	189.9	47	233.6
8	85.4	18	118.8	28	162.2	38	192.8	48	241.8
9	88.5	19	123.0	29	165.5	39	196.6	49	250.3
10	91.5	20	127.3	30	167.9	40	199.5	50	254.1

N°	Code	N°	Code	N°	Code	N°	Code	N°	Code
1	D023N	22	D131N	43	D251N	64	D371N	85	D532N
2	D025N	23	D132N	44	D252N	65	D411N	86	D546N
3	D026N	24	D134N	45	D255N	66	D412N	87	D565N
4	D031N	25	D143N	46	D261N	67	D413N	88	D606N
5	D032N	26	D145N	47	D263N	68	D423N	89	D612N
6	D036N	27	D152N	48	D265N	69	D431N	90	D624N
7	D043N	28	D155N	49	D266N	70	D432N	91	D627N
8	D047N	29	D156N	50	D271N	71	D445N	92	D631N
9	D051N	30	D162N	51	D274N	72	D446N	93	D632N
10	D053N	31	D165N	52	D306N	73	D452N	94	D645N
11	D054N	32	D172N	53	D311N	74	D454N	95	D654N
12	D065N	33	D174N	54	D315N	75	D455N	96	D662N
13	D071N	34	D205N	55	D325N	76	D462N	97	D664N
14	D072N	35	D212N	56	D331N	77	D464N	98	D703N
15	D073N	36	D223N	57	D332N	78	D465N	99	D712N
16	D074N	37	D225N	58	D343N	79	D466N	100	D723N
17	D114N	38	D226N	59	D346N	80	D503N	101	D731N
18	D115N	39	D243N	60	D351N	81	D506N	102	D732N
19	D116N	40	D244N	61	D356N	82	D516N	103	D734N
20	D122N	41	D245N	62	D364N	83	D523N	104	D743N
21	D125N	42	D246N	63	D365N	84	D526N	105	D754N

4
SETTING UP AND PROGRAMMING YOUR BAOFENG RADIO

Now that you understand the different frequency bands and their uses, it's time to set up and program your Baofeng Radio. In this chapter, we'll walk you through the process step-by-step, making it easy for even beginners to get up and running in no time.

First, let's start with setting up your Baofeng Radio. This involves charging the battery, installing the antenna, and turning on the radio. It's a simple process that can be completed in just a few minutes.

Next, we'll move on to programming your Baofeng Radio. This is where things can get a bit tricky, but don't worry, we'll make it easy for you. The first step is to locate the programming software for your specific model of the Baofeng Radio. This software can usually be found on the manufacturer's website or through a simple online search.

Once you have the programming software, it's time to enter the channels you want to use. Channels are simply specific frequencies that you can tune your radio to in order to communicate with others. The programming software will guide you through the process of entering the channels, and you can also refer to the user manual for more detailed instructions.

Finally, it's time to program your Baofeng Radio. This is where you'll enter the channels you want to use, as well as any other settings you want to customize, such as the volume, squelch, and tone settings.

Step-by-step guide on how to program your Baofeng Radio:

1. Locate the programming software:

- Start by searching for the programming software for your specific model of the Baofeng Radio on the manufacturer's website.
- Alternatively, you can search for the software online.

1. Install the software:

- Follow the instructions provided by the manufacturer to install the programming software on your computer.

1. Connect the radio to the computer:

- Connect your Baofeng Radio to your computer using the programming cable.
- Make sure the radio is turned off before connecting it to your computer.

1. Open the programming software:

- Once the software is installed, launch it on your computer.
- The programming software should recognize your Baofeng Radio once it's connected.

1. Enter the channels:

- Use the programming software to enter the channels you want to use.
- You can enter the channels one by one or import a pre-made channel list.
- Enter the frequency, tone, and other settings for each channel.
- Refer to the user manual for detailed instructions on entering the channels.

1. Save the channels:

- Once you've entered all the channels, save the programming data to your Baofeng Radio.
- Follow the instructions provided by the programming software to save the channels.

1. Disconnect the radio:

- Once the programming data has been saved, disconnect the Baofeng Radio from your computer.
- Turn on the radio to confirm that the channels have been saved and are working correctly.

Entering the channels is an important step in programming your Baofeng Radio. It allows you to tune in to specific frequencies, so you can communicate with others who are using the same frequencies. The channels you enter will depend on the specific needs and uses of your radio. You can enter frequencies for local amateur radio clubs, emergency services, or any other channels you want to use. By entering the channels, you ensure that your Baofeng Radio is always tuned in to the right frequencies, so you can communicate with others when you need to.

Once you've programmed your Baofeng Radio, it's time to hit the trails and start communicating!

In summary, setting up and programming your Baofeng Radio is a simple process that can be completed in just a few minutes. Whether you're a beginner or a seasoned amateur radio operator, the Baofeng Radio offers the flexibility and customization options you need to stay connected in the great outdoors. In the following chapters, we'll delve into more advanced programming techniques and provide tips and tricks for optimizing your radio's performance.

5
BASIC PROGRAMMING TECHNIQUES: A GUIDE TO PROGRAMMING CHANNELS AND FREQUENCIES ON THE RADIO

Programming your Baofeng Radio is essential for effective communication and customization to meet your specific needs.

The following guide provides a step-by-step explanation of the basic techniques required to program channels and frequencies on your radio.

1. Channel Programming:

- Locate the channel programming mode on your Baofeng Radio. This can typically be done by pressing and holding a specific button.
- Enter the channel number you want to program.

- Input the frequency you want to assign to that channel.
- Save the programming by pressing the appropriate button.

1. Frequency Programming:

- Locate the frequency programming mode on your Baofeng Radio. This can typically be done by pressing and holding a specific button.
- Input the frequency you want to program.
- Save the programming by pressing the appropriate button.

1. Importing Frequency Data:

- Some Baofeng Radios have the option to import frequency data from a computer.
- Connect your Baofeng Radio to your computer using a programming cable.
- Use the appropriate software to import the frequency data into your radio.

1. Scanning Frequencies:

- Your Baofeng Radio has the capability to scan for active frequencies.

- Locate the scanning mode on your radio and activate it.
- The radio will automatically scan through the programmed frequencies and stop any active ones.

By following these basic programming techniques, you can easily customize your Baofeng Radio to meet your specific needs and ensure effective communication. Whether you are a beginner or an experienced user, these techniques are valuable tools for maximizing the functionality of your radio.

6
STORING MULTIPLE CHANNELS: A GUIDE TO SAVING AND ORGANIZING CHANNELS FOR EASY ACCESS

As you program more channels and frequencies into your Baofeng Radio, it's important to have a system for organizing and accessing them quickly. In this chapter, we'll explore the best practices for storing multiple channels to ensure seamless communication and efficient access.

1. **Channel Banks:**

   - Most Baofeng Radios have the option to store channels into banks. This allows you to categorize and access your channels easily.
   - To store channels into banks, first, enter the channel programming mode on your radio.

- Next, select the channel you want to add to a bank.
- On most Baofeng Radios, you can then press a button to move the channel into a bank.
- For example, you may have one bank for personal channels, one for work channels, and another for emergency channels. To store a personal channel in the personal bank, you would select the channel and then press the button to move it into the personal bank.

1. **Favorites List:**

- Many Baofeng Radios have a favorites list that allows you to quickly access your most frequently used channels.
- To add a channel to your favorites list, first, enter the channel programming mode on your radio.
- Next, select the channel you want to add to your favorites list.
- On most Baofeng Radios, you can then press a button to add the channel to your favorites list.
- For example, if you use channel 123 frequently, you can add it to your favorites list so that you can quickly access it in the future.

1. **Alpha-Numeric Labeling:**

- Labeling your channels with descriptive names can make it easier to quickly identify and access them.
- To label a channel, first, enter the channel programming mode on your radio.
- Next, select the channel you want to label.
- On most Baofeng Radios, you can then enter a label for the channel using the keypad.
- For example, you could label channel 123 as "Home" or "Work" to quickly identify its purpose.

1. **Micro-SD Card Storage:**

- Some Baofeng Radios have the option to store channels and frequencies on a micro-SD card.
- To store channels and frequencies on a micro-SD card, first, insert the micro-SD card into the slot on your radio.
- Next, enter the channel programming mode on your radio.
- On most Baofeng Radios, you can then select the option to save the channels and frequencies to the micro-SD card.
- For example, if you have programmed several channels and frequencies into your radio, you can store them on a micro-SD card for safekeeping. This will allow you to easily

transfer your programming between radios and ensures that your channels are always accessible, even if you lose your radio.

By following these best practices for storing multiple channels, you can stay organized and ensure that you have quick access to the channels and frequencies you need. Whether you're a beginner or an experienced user, these techniques will help you maximize the functionality of your Baofeng Radio.

7
BATTERY MANAGEMENT: A GUIDE TO MAXIMIZING BATTERY LIFE AND CHARGING THE RADIO

Battery life is an important factor to consider when using a Baofeng radio. Proper battery management can help to ensure the radio is always ready to use when you need it. Here are some tips and procedures for maximizing battery life and charging the radio.

1. Use the right type of battery: Baofeng radios are designed to work with Lithium-Ion batteries. Using other types of batteries can lead to problems and could damage the radio.
2. Charge the battery regularly: Regular charging will help to keep the battery in good condition and extend its life. Avoid letting the battery run down completely before recharging it.

3. Use an appropriate charging method: Baofeng radios can be charged using the supplied charging dock or with a USB cable. If you are using the charging dock, ensure that it is connected to the correct power source and that the battery is properly seated in the dock. If you are using a USB cable, ensure that it is connected to a suitable USB power source.
4. Monitor battery life: The Baofeng radio has a battery indicator that shows the level of charge. Check the battery level regularly and recharge the battery as needed.
5. Store the battery correctly: When not in use, store the battery in a cool, dry place. Avoid exposing the battery to extreme temperatures, which can damage it and reduce its life.
6. Keep the battery contacts clean: Dirt and corrosion can affect the performance of the battery and reduce its life. Regularly clean the battery contacts with a clean, dry cloth to maintain good electrical contact.
7. Turn off the radio when not in use: Turning off the radio when it is not in use can help to conserve battery life. The radio will automatically enter a low-power state when it is turned off.

By following these tips and procedures, you can ensure that your Baofeng radio has the maximum possible battery life and is always ready to use when you need it.

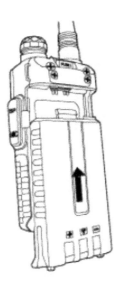

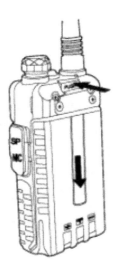

8
COMMUNICATION TIPS AND TRICKS: A GUIDE TO IMPROVING AUDIO QUALITY AND RELIABILITY

Radio communication is a vital tool in many situations and ensuring that you have a clear and reliable connection can mean the difference between life and death in certain scenarios. In this chapter, we'll cover some tips and tricks to help you maximize the quality of your radio communication and ensure that your message is heard loud and clear.

1. **Adjusting the Microphone:** Before you start communicating, it's crucial to adjust the microphone to your mouth. The microphone should be positioned about an inch away from your lips to avoid muffling your voice. Also, make sure the microphone is aimed directly at

your mouth, so your voice is transmitted clearly. Additionally, hold the speaker microphone in a stable position, away from your body and clothing, to minimize background noise.

2. **Talking in a Calm and Controlled Manner:** When communicating over the radio, it's essential to speak in a calm and controlled voice. Avoid speaking too fast, too slow, or too loud, as this can cause the signal to break up or become garbled. Take a deep breath, relax, and speak clearly and calmly into the microphone.

3. **Reducing Background Noise:** Background noise can be a major issue when communicating over the radio, so it's important to minimize it as much as possible. This can be done by closing windows, turning off other electronic devices, or moving to a quiet location. If you're communicating in a vehicle, close the windows, turn off the radio or other electronic devices that might be causing interference, or use noise-canceling headphones. Additionally, you can adjust the squelch setting on your radio, which helps to reduce background noise.

4. **Using a Good Antenna:** A good antenna is crucial to achieving clear and reliable

communication. Make sure you're using a high-quality, durable antenna that's designed for your radio. An antenna with a longer whip will typically provide a better performance, but it will also be more cumbersome to carry. Consider your specific needs and the environment you'll be using the radio when choosing an antenna.

5. **Antenna placement:** The proper placement of the antenna can significantly impact the audio quality and reliability of your radio communication. To get the best performance from your radio, make sure that the antenna is placed in an optimal location, away from metallic objects and electronic interference. For example, if you are using a handheld Baofeng radio, hold it vertically, with the antenna facing upwards, to receive the best signal.

6. **Volume control:** Proper volume control is crucial for ensuring clear communication on your Baofeng radio. Make sure that you set the volume at an appropriate level, so that your voice can be heard clearly by the receiving party, but not too loudly that it becomes distorted. Additionally, you can adjust the microphone gain to control the volume of your voice transmission.

By following these tips and tricks, you can improve the audio quality and reliability of your radio communication.

9
OPERATING IN OUTDOOR ENVIRONMENTS: A GUIDE TO USING THE RADIO IN REMOTE AND RURAL AREAS

As a Baofeng radio user, you may find yourself operating in remote or rural areas, where you need to rely on your radio for communication. In these situations, it is important to understand how to maximize your radio's performance and ensure reliable communication.

Read the following to help you get the most out of your Baofeng radio in outdoor environments:

1. **Understanding the Terrain:** Before venturing out into remote or rural areas, it's important to understand the terrain you'll be operating in. This includes researching the topography,

geography, and weather patterns of the area to determine how they may impact your ability to communicate effectively. For example, if you're operating in a mountainous area, you may need to take into account the impact of the mountains on your signal, or if you're operating in a forested area, you may need to consider the impact of trees on your signal.

2. **Antenna Selection:** Choosing the right antenna for your radio is critical for operating in remote and rural areas. Different types of antennas have different capabilities, and it's important to choose the one that will best meet your needs. For example, a directional antenna may provide a stronger signal in a particular direction, but it may also be more challenging to use than an omnidirectional antenna.

3. **Proper Power Management:** Maintaining adequate power levels is essential when operating in remote and rural areas. This means carrying extra batteries, a solar panel, or a generator if necessary. For example, if you're camping in a remote location, you may need to carry extra batteries or a solar panel to keep your radio charged.

4. **Use of Repeaters:** Repeaters are stations that receive and retransmit signals, and they can be a valuable tool for improving communication

in remote and rural areas. When choosing a repeater, it's important to consider the coverage area, frequency, and compatibility with your radio. For example, if you're operating in a remote location, you may be able to use a repeater to reach a wider area or to connect with other radios that are farther away.

5. **Operating in Harsh Conditions:** Your Baofeng radio is designed to withstand rugged conditions, but it is important to take proper care when operating in extreme temperatures, heavy rain, or other harsh environments. In this section, we'll provide advice on how to protect your radio and ensure reliable communication in these conditions.

6. **Emergency Communication:** When in remote or rural areas, it is crucial to be prepared for emergencies. In this section, we'll discuss the importance of creating an emergency communication plan and provide tips for using your Baofeng radio in emergency situations.

By following the tips and techniques outlined in this chapter, you'll be able to maximize your radio's performance and ensure reliable communication, even in the most challenging outdoor environments.

10
EMERGENCIES AND DISASTER PREPAREDNESS: A GUIDE TO USING THE RADIO IN EMERGENCY SITUATIONS

I n times of emergency, having a reliable means of communication is crucial. Whether it's a natural disaster, a power outage, or any other crisis, having a Baofeng radio can make a significant difference. In this chapter, we'll explore how to use the radio in emergency situations and provide tips for disaster preparedness.

1. **Understanding the Importance of Communication During Emergencies**

    - In times of emergency, having reliable and effective communication can be the difference between life and death. In a disaster scenario, traditional communication methods such as

landline phones or the internet may not be available, making the use of a radio a crucial tool for staying informed and staying in touch with loved ones.
- A practical example of this can be seen during Hurricane Katrina in 2005. People who were prepared with a radio were able to listen to real-time updates from emergency services, whereas those without a radio were left in the dark, not knowing the extent of the damage or where to seek help.

1. **Choosing the Right Radio for Emergency Situations**

- In emergency situations, it is important to have a radio that is reliable, durable, and easy to use. When choosing a radio, it's essential to consider factors such as battery life, the range of frequencies it can pick up, and its durability in various weather conditions.
- A practical example of choosing the right radio can be seen when hiking in a remote area. A hiker who brings a radio with a long battery life and weather resistance is better prepared in case of an emergency, as they can use it to call for help or stay updated on the weather conditions.

1. **Staying Informed During Emergencies with the Radio**

- In emergency situations, it is crucial to stay informed about the latest developments. Using the radio, you can tune into emergency broadcasts and updates from emergency services to stay informed about the situation.
- A practical example of staying informed during an emergency can be seen during a forest fire. By tuning into emergency broadcasts on the radio, a person can stay updated on the extent of the fire, where it is spreading, and what evacuation measures are in place.

1. **Establishing Communication During Emergencies with the Radio**

- In an emergency situation, it's essential to stay in touch with loved ones and emergency services. With the radio, you can establish two-way communication and let others know your situation and needs.
- A practical example of establishing communication during an emergency can be seen during a natural disaster such as an earthquake. By using the radio, a person can communicate with emergency services to

report the extent of the damage and receive instructions on what to do next. They can also communicate with loved ones to let them know they are safe.

In conclusion, having a radio during emergencies and disaster scenarios can be a lifesaver. From staying informed about the latest developments to establishing communication with loved ones and emergency services, the radio is a crucial tool for staying safe in emergency situations.

11
GUERRILLA WARFARE TECHNIQUES: A GUIDE TO USING THE RADIO FOR MILITARY AND TACTICAL OPERATIONS

The use of radios in military and tactical operations can mean the difference between success and failure. Guerrilla warfare tactics, which rely heavily on the element of surprise, can be greatly aided by effective communication. In this chapter, we will explore several key techniques for using radios in guerrilla warfare operations.

1. **Establishing a secure communication network:** In guerrilla warfare, it's crucial to establish a secure communication network that can be used to coordinate actions and share intelligence. This can be accomplished by using encryption methods and secure frequencies

that are less likely to be intercepted by the enemy. When setting up the network, it's also important to consider factors such as the range of the radios and the number of people that need to be included in the network.

2. **Using code language:** Using code language, also known as radio shorthand or brevity codes, can help to keep communications secure and reduce the risk of interception. Brevity codes are a set of standardized phrases that are used to convey information quickly and succinctly. For example, the phrase "contact front" might mean that enemy forces have been sighted.

3. **Deploying covert communication techniques:** In guerrilla warfare operations, it's often necessary to communicate covertly in order to avoid detection by the enemy. This can be accomplished by using techniques such as morse code, low-power transmissions, or transmitting on frequencies that are not being monitored by the enemy. Additionally, it may be necessary to change frequencies frequently in order to stay ahead of enemy efforts to intercept the communication.

4. **Maintaining radio discipline:** Maintaining radio discipline is crucial in guerrilla warfare

operations. This means limiting transmission time, avoiding unnecessary chatter, and adhering to strict procedures for communicating information. By maintaining radio discipline, guerrilla forces can reduce the risk of giving away their position and ensure that vital information is not lost due to interference or other factors.

By mastering these guerrilla warfare techniques, military and tactical operators can effectively use their radios to coordinate actions, share intelligence, and maintain the element of surprise. With the ability to communicate effectively, guerrilla forces can increase their chances of success and achieve their mission objectives.

CAMPING PROCEDURES FOR GUERRILLA UNITS

Establishing camps for guerrilla units boosts motivation and reduces distractions, improving the cooperative spirit among small units by linking the physical environment with the psychological one. The squad leader must implement regular camping procedures, starting with choosing the right location for the camp. The leader should select a site that offers two to three escape routes and is dominant in the area. He will then assign tasks to members of the unit, such as cleaning the camping area, providing proper drainage and trenches in case of rain or emergency, constructing a stove and a windbreaking wall, setting up a latrine and waste disposal area, and posting a watchman and setting the password. The leader must also establish alternate meeting points in case of a hasty retreat.

These procedures enhance motivation and strengthen the cooperative spirit in the unit, as they provide a sense of security, order, and belonging that are crucial for maintaining the guerrilla's morale. In addition to good physical conditions, regular group discussions and self-criticism sessions are essential for promoting good psychological conditions. The act of breaking camp through collective effort reinforces the unity of thought and strengthens the group's spirit.

Interaction with the Community

To secure popular support, crucial for successful guerrilla warfare, leaders must encourage positive interaction between guerrillas and civilians by adopting the principle of "living, eating, and working with the people" and controlling their activities. Group discussions should focus on positive identification with the people and avoid discussing military tactics. The Communist threat should be highlighted as the primary enemy of the people and a secondary threat to the guerrilla forces. Guerrilla members with high political awareness and disciplined conduct should be sent to populated areas to engage in armed propaganda and promote the principles of respecting human rights and property, helping with community work, protecting the people from Communist aggression, and teaching hygiene and education to win the people's trust and increase their democratic preparedness. This approach will foster sympathy and support from the peasants, who will provide logistical support, intelligence information, and even participate in combat. Guerrilla members should persuade through words, not weapons, to earn respect and increase the acceptance of their message.

When conducting tactical operations in populated areas, guerrillas must also carry out parallel psychological actions to explain their objectives, which are to bring peace, liberty, and democracy to all people without exception, and that their struggle is not against the nationals but against other countries imperialism. This approach will enhance psychological achievements and improve future operations.

In guerrilla warfare, the face-to-face persuasion of the

guerrilla and the community plays a crucial role in psychological operations. The leaders must implement effective camping procedures, promote positive interaction with the community, and engage in armed propaganda to increase motivation, strengthen the cooperative spirit, and secure popular support.

UNDERSTANDING THE DIFFERENT MODES OF OPERATION: A GUIDE TO THE DIFFERENT OPERATING MODES ON THE RADIO

Radio communication systems are designed to operate in different modes, each with its own set of specific characteristics and limitations. The mode of operation is determined by the type of modulation used, the frequency band, the power output, and the number of users. In this chapter, we'll take a detailed look at the different modes of operation used in radio communication and what each mode is best suited for.

1. **Amplitude Modulation (AM):** Amplitude Modulation (AM) is a type of modulation in which the amplitude of the carrier wave is varied in proportion to the modulating signal. The modulating signal can be audio, digital, or

other types of signals. AM is commonly used in broadcast radio and is the oldest form of modulation.
2. **Frequency Modulation (FM):** Frequency Modulation (FM) is a type of modulation in which the frequency of the carrier wave is varied in proportion to the modulating signal. The modulating signal can be audio, digital, or other types of signals. FM is commonly used in radio communications for voice communication. It is more immune to noise and interference than AM and provides better audio quality.
3. **Single-Sideband (SSB):** Single-Sideband (SSB) is a type of modulation in which only one sideband of the carrier wave is transmitted, instead of the full carrier wave. This reduces the bandwidth required for transmission and is commonly used in long-range communications, such as amateur radio and maritime communications.
4. **Continuous Wave (CW):** Continuous Wave (CW) is a type of modulation in which a continuous unmodulated carrier wave is transmitted. It is often used for Morse code transmissions, and its simple signal structure makes it very easy to detect and decode.

5. **Digital Modes:** Digital modes are a type of modulation in which digital signals are transmitted over the airwaves. Digital modes can be more efficient and resistant to noise and interference than analog modes. They can also provide more data transmission capability. Some examples of digital modes are:

- Packet Radio: A method of transmitting digital data using short packets of information. It is commonly used in amateur radio and other short-range communications.
- Digital Mobile Radio (DMR): A digital voice and data transmission protocol used in professional mobile radio communications.
- Automatic Packet Reporting System (APRS): A digital communications system that uses packet radio to transmit real-time position, weather, and other data.

1. **Spread Spectrum:** Spread Spectrum is a type of modulation in which the frequency of the carrier wave is spread over a wide bandwidth. This makes it difficult for an eavesdropper to intercept the signal and provides improved resistance to interference. Spread Spectrum is commonly used in military and industrial

communications, as well as in wireless local area networks (WLANs).

AM (Amplitude Modulation) radio frequencies are typically found in the range of 550 to 1700 kilohertz (kHz). Some common AM radio stations include:

- 880 kHz: WCBS, New York
- 1080 kHz: WKXW, Trenton, New Jersey
- 1230 kHz: KDOW, San Francisco Bay Area

FM (Frequency Modulation) radio frequencies are typically found in the range of 87.5 to 108 megahertz (MHz). Some common FM radio stations include:

- 88.1 MHz: WKSU, Kent, Ohio
- 91.1 MHz: WBEZ, Chicago
- 94.5 MHz: KFOG, San Francisco

It's important to note that these frequency ranges and specific station examples may vary depending on your location and local regulations. However, these examples can give you a general idea of the typical frequency ranges for AM and FM radio.

1. **SSB (Single Sideband):** SSB is commonly used for voice communication on amateur radio

bands, and it can be found in **the frequency range of 3-30 MHz. A common frequency for SSB voice communication on amateur radio bands is 14.2 MHz.**

2. **CW (Continuous Wave):** CW, <u>also known as Morse code</u>, is primarily used for text communication on amateur radio bands. It can be found in the **frequency range of 1.8-30 MHz. A common frequency for CW communication on amateur radio bands is 14.0 MHz.**

3. **Digital Modes:** Digital modes, such as **PSK31, RTTY, and JT65**, are <u>used for digital data communication on amateur radio bands</u>. They can be found in the **frequency range of 1.8-30 MHz, with a common frequency for digital mode communication being 14.1 MHz.**

Note: These are just examples, and the actual frequencies may vary depending on the region and the band in use. Additionally, these frequencies may be subject to change as the amateur radio community evolves and new modes of operation are developed.

. . .

In conclusion, there are a variety of modes of operation used in radio communication, each with its own set of specific characteristics and limitations. The mode of operation is determined by the type of modulation used, the frequency band, the power output, and the number of users. Understanding the different modes of operation is important for selecting the best mode for your communication needs and for operating your radio system effectively and efficiently.

13
"ANTENNA SELECTION AND CONFIGURATION: A GUIDE TO CHOOSING AND CONFIGURING THE RIGHT ANTENNA FOR THE RADIO"

Radio communication requires a proper antenna system in order to work effectively. Choosing the right antenna for your radio system is a crucial aspect of radio communication, as it can greatly impact the range and quality of your transmissions. This chapter will guide you through the different factors to consider when selecting and configuring your radio antenna.

ANTENNA TYPES

There are a variety of antenna types to choose from, each with its own advantages and disadvantages.

. . .

Some of the most commonly used antenna types include:

- **Dipole Antennas:** Dipole antennas are simple and inexpensive and are often used as a starting point for new radio operators. They are easy to install but have a limited range and are highly directional.
- **Yagi Antennas:** Yagi antennas are directional and provide good gain, making them suitable for long-range communication. They consist of multiple elements, which need to be carefully aligned in order to provide maximum performance.
- **Ground Plane Antennas:** Ground plane antennas are also directional and provide good gain but are less complex than Yagi antennas. They consist of a single element that is grounded and is commonly used for mobile radio applications.
- **Loop Antennas:** Loop antennas are circular or square-shaped antennas that are directional and provide good gain, making them suitable for long-range communication. They are less commonly used due to their complex design and construction requirements.

ANTENNA CONFIGURATION

Once you have chosen the right antenna for your radio system, you will need to configure it correctly.

Some of the key factors to consider when configuring your antenna include:

- Height: The height of your antenna will impact the range and quality of your transmissions. A higher antenna will generally provide a better range and quality than a lower antenna.
- Orientation: The orientation of your antenna will also impact the range and quality of your transmissions. A horizontally oriented antenna will generally provide better range, while a vertically oriented antenna will provide better quality.
- Ground Plane: The ground plane of your antenna is an important factor to consider, as it can greatly impact the performance of your antenna. A large, flat, and conductive ground plane will generally provide better performance than a smaller or irregularly shaped ground plane.
- Matching: Proper matching between the antenna and your radio is crucial, as it will impact the range and quality of your

transmissions. There are a variety of matching techniques that can be used, including impedance matching, frequency matching, and power matching.

ANTENNA MAINTENANCE

In order to ensure optimal performance, it is important to regularly maintain your antenna system. Some of the key maintenance tasks to consider include:

- Cleaning: Regular cleaning of your antenna will help to remove any debris or buildup that can negatively impact performance.
- Inspection: Regular inspection of your antenna will help to identify any potential issues, such as damage or corrosion, that may need to be repaired or replaced.
- Tuning: Regular tuning of your antenna system will help to ensure optimal performance and range and may also be required if you have made any changes to your radio system.

In conclusion, selecting and configuring the right antenna for your radio system is a crucial aspect of radio communication. By carefully considering the different types of antennas available, as well as the configuration factors, antenna maintenance tasks, and proper matching techniques, you can ensure optimal performance and range for your radio transmissions.

14
TROUBLESHOOTING AND MAINTENANCE: A GUIDE TO DIAGNOSING AND FIXING COMMON PROBLEMS WITH THE RADIO

I n this chapter, we will cover common problems that may occur with radio and how to diagnose and fix them. Proper maintenance is crucial to ensuring that your radio operates at peak performance and avoiding costly repairs.

1. **No Transmit or Receive Audio:** One of the most common problems in radio communication is the lack of transmit or receive audio. This can be caused by several factors, including incorrect microphone or speaker connections, faulty cables, or a problem with the internal audio circuits. To troubleshoot this problem, first check the

connections and cables, making sure they are securely connected and in good condition. If the cables are fine, then try adjusting the audio levels in the radio's menu system. If the problem persists, you may need to take the radio to a professional repair service.

2. **Intermittent Reception:** Intermittent reception can be frustrating and is usually caused by poor antenna placement, a damaged antenna, or a weak signal. To troubleshoot this problem, first check the antenna placement, making sure it is positioned in a clear and unobstructed area. If the antenna is damaged, replace it with a new one. If the problem persists, you may need to consider upgrading your antenna to a higher-gain antenna or placing the antenna in a more favorable location.

3. **Low Transmit Power:** If you're experiencing low transmit power, it could be due to a weak battery, incorrect settings, or a malfunctioning power amplifier. To troubleshoot this problem, first, check the battery levels and make sure it's fully charged. If the battery is fine, then check the radio's menu system for incorrect power settings. If the problem persists, you may need to take the radio to a professional repair service.

4. **Blowing Fuses:** Blowing fuses are a common problem in radio communication, and they can be caused by a short circuit, incorrect wiring, or a malfunctioning component. To troubleshoot this problem, first, check the wiring, making sure it's correctly connected and in good condition. If the wiring is fine, then replace the fuse with a new one. If the problem persists, you may need to take the radio to a professional repair service.

5. **Distorted Audio:** Distorted audio is another common problem in radio communication, and it can be caused by overloading the audio circuits, incorrect settings, or a malfunctioning component. To troubleshoot this problem, first, check the audio levels in the radio's menu system and make sure they are set correctly. If the settings are fine, then check the microphone and speaker connections, making sure they are securely connected and in good condition. If the problem persists, you may need to take the radio to a professional repair service.

6. **Poor Receiver Sensitivity:** Poor receiver sensitivity can be caused by several factors, including weak signals, incorrect settings, or malfunctioning components. To troubleshoot this problem, first check the radio's menu

system for incorrect settings, such as the RF gain or squelch settings. If the settings are fine, then check the antenna placement and make sure it's positioned in a clear and unobstructed area. If the problem persists, you may need to consider upgrading your antenna or taking the radio to a professional repair service.

In conclusion, the troubleshooting and maintenance chapter is a comprehensive guide that covers the most common problems that may occur with the radio. By following the procedures listed in this chapter, radio operators can diagnose and fix problems quickly and effectively, ensuring that their radio is always in good working condition.

15
"ACCESSORIES AND ADD-ONS: A GUIDE TO THE DIFFERENT ACCESSORIES AND ADD-ONS AVAILABLE FOR THE RADIO"

The use of accessories and add-ons can greatly enhance the capabilities and performance of a radio. In this chapter, we will explore the different types of accessories and add-ons that are available for radio, and how they can be used to improve the user's experience.

1. Antenna Tuners: Antenna tuners are accessories that help match the impedance of an antenna to the transmitter, allowing for maximum power transfer and better signal quality. Antenna tuners can be manual or automatic and are especially useful for those operating on multiple frequencies, or for those

using antennas that have a non-ideal impedance match to the transmitter.
2. Microphones: Microphones are an essential accessory for those using voice modes on their radio, such as SSB or FM. Microphones come in a variety of styles, including hand-held, desktop, and boom-mounted, and can be either dynamic or condenser types. When choosing a microphone, it's important to consider factors such as the frequency response, noise level, and sensitivity, as these will all impact the quality of the audio transmitted.
3. Power Supplies: Power supplies are an essential accessory for those who plan to operate their radio for extended periods of time, such as during field operations or contests. Power supplies can be either AC or DC and can be configured for either continuous operation or battery backup. When selecting a power supply, it's important to consider the voltage and current requirements of the radio, as well as the level of regulation required for stable operation.
4. Antenna Switches: Antenna switches are accessories that allow the user to quickly switch between multiple antennas, allowing for improved versatility and flexibility when operating. Antenna switches can be manual or

automatic and can be configured for either coaxial or balanced-line connections. When selecting an antenna switch, it's important to consider the number of antennas that will be used, as well as the type of connections required.

5. Headphones: Headphones are an accessory that can be used for listening to the audio from a radio, either during normal operation or for monitoring purposes. Headphones come in a variety of styles, including over-the-ear, in-ear, and headset-style, and can be either open-back or closed-back. When selecting headphones, it's important to consider the frequency response, noise isolation, and comfort, as these will all impact the listening experience.

6. Digital Interfaces: Digital interfaces are accessories that allow the user to interface their radio with a computer, enabling the use of digital modes such as PSK31, RTTY, and JT65. Digital interfaces can be either USB or serial-based and can be used for both audio input and output, as well as for control functions. When selecting a digital interface, it's important to consider the compatibility with the radio, as well as the software and operating system being used.

In conclusion, the use of accessories and add-ons can greatly enhance the capabilities and performance of a radio. Whether it's for improved audio quality, extended operating time, or increased versatility, the right accessory can make all the difference. As always, it's important to carefully consider the requirements and goals of the user before making a purchase, as this will ensure that the right accessory is chosen for the job.

CONCLUSION AND FUTURE OUTLOOK: A SUMMARY OF THE KEY TAKEAWAYS AND FUTURE DEVELOPMENTS IN THE RADIO INDUSTRY

In this final chapter, we aim to provide a comprehensive summary of all the key takeaways from the previous chapters and highlight future developments in the radio industry. The radio industry has come a long way since its invention, and with the ever-evolving technology, it is continuously changing and adapting to the modern world.

1. **Key Takeaways:**

- The radio industry has a rich history, from its invention to the various modes of operation and accessories available today.
- Understanding the different modes of operation, such as AM, FM, SSB, CW, and

digital modes, is crucial for effective communication using a radio.
- Antenna selection and configuration is also a crucial aspect in achieving optimal performance from the radio.
- Regular troubleshooting and maintenance can help extend the lifespan of the radio and prevent problems in the future.
- The radio industry has a vast range of accessories and add-ons available, from battery packs to external speakers, that can enhance the user's experience.
- In conclusion, the radio industry is constantly evolving, and it's important to stay informed and updated on the latest developments and advancements.

1. **Future Developments:**

- The rise of digital technologies has dramatically changed the radio industry, with digital radios replacing analog radios in many applications.
- There has also been a shift towards mobile and handheld radios, which are more portable and convenient for users on the move.
- The development of software-defined radios has opened up new possibilities for the radio

industry, allowing for greater flexibility and customization in radio communications.
- The integration of artificial intelligence and machine learning in radio technology is also expected to be a major development in the future, enabling radios to better adapt to changing environments and provide more efficient communication.
- The radio industry is also working towards making radios more eco-friendly, by reducing their power consumption and using sustainable materials.

In conclusion, the radio industry has a rich history and an exciting future ahead. With new technologies and advancements continuously being developed, the possibilities for the radio industry are endless. It's important to stay informed and updated on these developments to ensure that we can make the most of our radio equipment and achieve effective communication.

FINAL THANKS

Dear reader, I sincerely thank you for choosing to purchase this book. I hope you found reading interesting and fulfilling. As a writer, I know how important it is to get feedback on your writing. As a self-published independent writer, I kindly ask you to leave a positive and honest review of your book on the online retail site where you purchased it. Reviews are essential for independent writers like me, as they help us reach new readers and broaden our audience base. I hope you enjoyed your reading experience and that we can meet again between the pages of another book. Thanks again for your support.